Isaac Newton

VIP
Very Interesting People

Bite-sized biographies of Britain's most fascinating historical figures

CURRENTLY AVAILABLE

1. **William Shakespeare**	*Peter Holland*
2. **George Eliot**	*Rosemary Ashton*
3. **Charles Dickens**	*Michael Slater*
4. **Charles Darwin**	*Adrian Desmond, James Moore, and Janet Browne*
5. **Isaac Newton**	*Richard S. Westfall*
6. **Elizabeth I**	*Patrick Collinson*
7. **George III**	*John Cannon*
8. **Benjamin Disraeli**	*Jonathan Parry*
9. **Christopher Wren**	*Kerry Downes*
10. **John Ruskin**	*Robert Hewison*

FORTHCOMING

11. **James Joyce**	*Bruce Stewart*
12. **John Milton**	*Gordon Campbell*
13. **Jane Austen**	*Marilyn Butler*
14. **Henry VIII**	*E. W. Ives*
15. **Queen Victoria**	*H. C. G. Matthew and K. D. Reynolds*
16. **Winston Churchill**	*Paul Addison*
17. **Oliver Cromwell**	*John Morrill*
18. **Thomas Paine**	*Mark Philp*
19. **J. M. W. Turner**	*Luke Herrmann*
20. **William and Mary**	*Tony Claydon and W. A. Speck*

Isaac Newton

VIP

Very Interesting People

Richard S. Westfall

OXFORD
UNIVERSITY PRESS

OXFORD
UNIVERSITY PRESS

Great Clarendon Street, Oxford ox2 6DP

Oxford University Press is a department of the University of Oxford.
It furthers the University's objective of excellence in research, scholarship,
and education by publishing worldwide in

Oxford New York

Auckland Cape Town Dar es Salaam Hong Kong Karachi
Kuala Lumpur Madrid Melbourne Mexico City Nairobi
New Delhi Shanghai Taipei Toronto

With offices in

Argentina Austria Brazil Chile Czech Republic France Greece
Guatemala Hungary Italy Japan Poland Portugal Singapore
South Korea Switzerland Thailand Turkey Ukraine Vietnam

Oxford is a registered trade mark of Oxford University Press
in the UK and in certain other countries

Published in the United States
by Oxford University Press Inc., New York

First published in the *Oxford Dictionary of National Biography* 2004
This paperback edition first published 2007

First published 2007

British Library Cataloguing in Publication Data

Data available

Library of Congress Cataloging in Publication Data

Data available

Typeset by SPI Publisher Services, Pondicherry, India
Printed in Great Britain
on acid-free paper by
Clays Ltd, St Ives plc

ISBN 978-0-19-953452-4

1

Contents

Abbreviations

CUL	Cambridge University Library
King's Cam.	King's College, Cambridge
MIT Cam.	Massachusetts Institute of Technology, Cambridge

About the author

Richard S. Westfall was Professor of the History and Philosophy of Science at Indiana University, Bloomington. His books include *Science and Religion in Seventeenth-Century England*, *Force in Newton's Physics: the Science of Dynamics in the Seventeenth Century*, and *Never at Rest: A Biography of Isaac Newton*.

1

Beginnings in science

Sir Isaac Newton (1642–1727), natural philosopher and mathematician, was born on 25 December 1642 in the manor house of Woolsthorpe, near Colsterworth, about 7 miles south of Grantham, Lincolnshire, the only and posthumous son of Isaac Newton (1606–1642), yeoman farmer, and his wife, Hannah (*c*.1610–1679), daughter of James Ayscough, gentleman, of Market Overton, Rutland. His family could trace its roots to John Newton of Westby, a village near Grantham; John Newton's grandfather Simon Newton was a husbandman whose name appeared among taxpayers listed in Westby in 1524. The family clearly had industry and ability, each generation inching up the local hierarchy. Thus the son of John Newton of Westby styled himself yeoman in his will of 1562, a step above husbandman, and successive wills indicate that the family flourished until its members could be counted among the most prosperous farmers in the district. They were also prolific, so that the area south of Grantham came to be sprinkled with numerous thriving Newton families, descended from Simon Newton of Westby.

A scholarly child

Isaac Newton's own endowment was secure. One of his ancestors had purchased a farm in Woolsthorpe for a son, Richard; the manor of Woolsthorpe was added in 1623, and the combined estate was inherited by Newton's father in 1641. His mother, Hannah, brought a property worth £50 per annum to the marriage—her brother William occupied the rectory of Burton Coggles, two miles east of Woolsthorpe. As soon as the estate was settled Newton's parents married, in April 1642, but six months later, early in October, his father died, leaving a pregnant widow. For a yeoman, he also left a considerable estate, which would bear upon the prospects of his unborn son. In addition to extensive lands, there were goods and chattels valued at £459, including a flock of 235 sheep and 46 head of cattle, possessions in excess of the average at that time. Although it is impossible to assess accurately, an annual value of £150 appears to be a reasonable estimate of the estate.

Isaac Newton was born on Christmas day, a tiny baby. He was not expected to survive but his mother delayed his baptism until 1 January 1643. The next recorded event in his life was more than three years later, when his mother married the Revd Barnabas Smith, left her first-born son with her parents in Woolsthorpe, and went off to the rectory of North Witham, 1½ miles south of Woolsthorpe, to raise a second family. The Revd Smith was wealthy, with an estate worth £500 per annum in addition to the rectory. There is some evidence that Newton, who never mentioned his stepfather, hated him, and he probably did not care for his grandparents either. He recorded no affectionate recollection of his grandmother, and made no mention whatever of his grandfather. The animosity was shared; grandfather Ayscough pointedly omitted Isaac from his will. The unpleasant aspects of Newton's mature personality may have had their roots in this traumatic period.

The family were reunited in 1653, when Smith died and Newton's mother returned to Woolsthorpe with his half-brother and two half-sisters. By this time Newton's education had begun in day schools in neighbouring villages, undoubtedly thanks to his Ayscough upbringing. At least in William Ayscough, who held an MA from Cambridge, and probably in other family members, the Ayscoughs had a history of education, whereas neither Newton's father nor his uncles on the paternal side were able to sign their names. His education continued when, less than two years after the return of his mother to Woolsthorpe, he was sent off to the free grammar school of King Edward VI in Grantham.

In Grantham, Newton lodged with an apothecary, a Mr Clark. About his studies very little is known but it may be assumed that the master, a Mr Stokes, did what nearly every schoolmaster of the age did—that is, he concentrated on building a mastery of Latin. At any rate, about the middle of his undergraduate career, Newton began to compose as readily in Latin as in English. Although mathematics does not figure in the Newton papers from this time, a recent discovery has indicated that Stokes had more than ordinary mathematical competence, and he may have played some role in directing Newton down this path. However, the stories that survived in Grantham had to do with other things: building models, including a working model of a windmill, constructing sundials, and frightening the towns-people with a lantern attached to a kite. There was apparently a schoolboy romance with the stepdaughter of his landlord (though hers is the only account of it). Years later, the young woman, then Mrs Vincent, was one of two people from Grantham that Newton recalled with pleasure.

Late in 1659, as he neared the age of seventeen, Newton's mother called him home from school to take charge of her considerable estate. The experiment was a disaster. Newton wanted none of it. He had discovered the world of learning, and there his heart lay. Set to watch the sheep, he would, it was said, build models of water-wheels while the sheep wandered into the neighbours' fields, leaving his mother to pay for the damage. Stokes had assessed his student's potential and offered to remit the fee if Newton were to return, and the Revd William Ayscough likewise urged his sister to let the young man prepare for the university. In the autumn of 1660 she relented. Newton returned to Grantham, and nine months later he departed for Cambridge University.

To Cambridge

Newton arrived in Cambridge on 4 June 1661 and the following day presented himself at Trinity College, where he was entered in the books as a sub-sizar. Sizars and sub-sizars—the college does not appear to have made a distinction between them—were poor students who earned their keep by performing menial tasks. Although less segregated in Trinity than in other colleges, they were nevertheless a lower order who, for example, were not allowed to dine with their fellow students.

Newton's status as sizar raises questions. Heir to the lordship of a manor, son of a widow who can only be described as wealthy, he was used to being served, not to serving. Two explanations that are not mutually exclusive present themselves. His mother had begrudged him a full grammar school education, and she may have been unwilling to send him to the university on any other terms (his patchy undergraduate accounts that survive seem to suggest a stingy allowance). Alternatively he may have gone to Trinity as the sizar of Humphrey Babington, a fellow of the college who had Lincolnshire connections, whose sister was the wife of the apothecary in Grantham with whom Newton had stayed. Babington held the rectory of Boothby Pagnell, not far from Woolsthorpe; Newton stayed at Boothby Pagnell for part of the time he was home during the plague, and a number of other details connect the two men. Babington was resident in Trinity College only four or five weeks of the year and may have needed someone to watch over his affairs.

In any event it is safe to assume that Newton's lowly status as a sub-sizar served to magnify his isolation from other students. He formed a friendship with one, John Wickins, who became his chamber fellow and, like Newton, stayed on as a fellow of the college. Although a number of others who were students with Newton also became fellows and lived in the college with him for thirty-five years, there is no evidence from either side of any relationships between them. His independence as a student, extending a pattern already manifest in grammar school, formed the initial chapter of an isolation within the university that continued until he left. After his departure he never corresponded with anyone he had known there.

Evidence of Newton's studies comes from the reading notes that he kept, especially those in one notebook. They show that his tutor, Benjamin Pulleyn, set him a curriculum the roots of which stretched back four centuries to the establishment of the medieval universities. Its focus was Aristotle: Newton read Aristotelian logic, Aristotelian ethics, and Aristotelian rhetoric, all preparations for the study of Aristotelian philosophy. Johannes Magirus's *Physiologiae peripateticae* furnished his first serious introduction to natural philosophy—Aristotelian natural philosophy, of course, as the name of the book implies.

In what became his characteristic style, Newton made notes from both ends of the notebook. It is significant that he did not finish reading any one of the texts he was set, and suddenly, apparently early in 1664, although nothing fixes the date with certainty, he broke off his reading in the established curriculum to pursue a different line of study. There is no hint of tutorial guidance. By every indication, Newton launched his new course alone. In the notebook some 200 pages in the middle remained empty. On one of them he entered a title, 'Questiones quaedam philosophcae' [*sic*], and he set down forty-five headings under which to gather information from his new reading (CUL, Add. MS 3996, fols. 88–135). Later he copied a slogan under the title: 'Amicus Plato amicus Aristoteles magis amica veritas'. Neither Plato nor Aristotle appeared again in the 'Quaestiones'. What did appear, significantly, were notes from Descartes, then reported in Cambridge, as in Paris, as the leading exponent of a revolutionary new philosophy. Newton digested Descartes's writings in a way he had not done the readings prescribed by the curriculum. He also took notes from Walter Charleton's English digest of Pierre Gassendi, from Galileo's *Dialogue*, and from the writings of Robert Boyle, Thomas Hobbes, Kenelm Digby, Joseph Glanville, and Henry More, all in their way exponents of unauthorized philosophical positions, and interested in the new science.

The 'Quaestiones quaedam' launched Newton's career in science. They set out his initial interest in a number of key phenomena, such as the cohesion of bodies, capillary action, the expansion of gases, and surface tension, that continued to occupy central places in his ongoing speculations on the nature of things. Also noted was his first concern with the question of colours, the problem of motion, and the cause of gravity in terrestrial bodies. More than anything else, the 'Quaestiones' considered the relative merits of the Cartesian and the Gassendist (that is, atomist) versions of the mechanical philosophy, and began to lean towards the latter.

'My age for invention'

2

Anni mirabiles, 1665–1666

One of the central stories in the Newton legend concerns the *annus mirabilis*, the marvellous year of discovery to which all of his later achievements in science could be traced. Newton himself helped to inaugurate the story in an often quoted passage that dates from much later:

> In the beginning of the year 1665 I found the Method of approximating series. ... The same year in May I found the method of Tangents of Gregory & Slusius, & in November had the direct method of fluxions & the next year in January had the Theory of Colours & in May following I had entrance into ye inverse method of fluxions.

Nor were his discoveries confined to mathematics. Newton went on to claim that:

> the same year I began to think of gravity extending to y[e] orb of the Moon & (having found out how to estimate the force with w[ch] [a] globe revolving within a sphere presses the surface of the sphere) from Keplers rule of the periodical times of the Planets being in sesquialterate proportion of their distances from the center of their Orbs, I deduced that the forces w[ch] keep the Planets in their Orbs must [be] reciprocally as the squares of their distances from the centers about w[ch] they revolve: & thereby compared the force requisite to keep the Moon in her Orb with the force of gravity at the surface of the earth, & found them answer pretty nearly. All this was in the two plague years of 1665–1666. For in those days I was in the prime of my age for invention & minded Mathematicks & Philosophy more than at any time since. (CUL, Add. MS 3968.41, fol. 85)

As Newton's words make clear, the legend should at least speak of *anni mirabiles*, in the plural, yet the historical evidence shows that much of the work mentioned was less complete than his words imply—it is now generally agreed that only with hindsight can the bulk of Newtonian science be traced to this early date. Never mind: the achievement of these years, measured by any standard other than that of the Newton legend, remains marvellous in the full sense of the word.

In 1664 mathematics seized Newton's attention, and in mathematics the marvellous years were most fully realized. When the surviving record of mathematical study during the following two years is considered it is hard to believe that he had time for anything else. By his own account, which his niece's husband, John Conduitt, wrote down, he was self-taught in mathematics:

> He bought Descartes's Geometry & read it by himself... when he was got over 2 or 3 pages he could understand no farther than he began again & got 3 or 4 pages farther till he came to another difficult place, than he began again & advanced farther & continued so doing till he made himself Master of the whole without having the least light or instruction from any body. (King's Cam., Keynes MS 130.10, fol. 2*v*)

With Descartes's *Geometry*, Newton studied the wealth of commentaries included in the second Latin edition. He also absorbed the Dutchman Franz van Schooten's *Miscellanies*, William Oughtred's *Clavis*, the works of François Viète on algebra, and the writings of John Wallis on infinitesimals. Within a year, working on his own, Newton had made himself the master of the new analysis created by earlier mathematicians of the seventeenth century. Gradually his reading notes transformed themselves into original investigations, and he launched forth on the sea of mathematics as an independent explorer.

The central mathematical discoveries

Newton's first major discovery came in quadratures, or what is now called integration. Following a method he found in Wallis, he set out to determine the area, in the first quadrant, of a segment under a circle. From earlier analysts he knew how to evaluate areas under curves of the simple power series such as $y = x, x^2, x^3 \ldots$, and Wallis had also taught him that the pattern observed in the results applies equally to fractional and negative exponents, and that with polynomials the area is the sum of areas under the separate terms, each computed according to the same rule. But what was one to do with curves such as $y = (1 - x^2)^{1/2}$, the circle, or $y = 1/x$, the equilateral hyperbola? Expanding and improving on Wallis's method, Newton arrived at the binomial theorem which expresses such areas as infinite series. In his enthusiasm for his new discovery he utilized the area under the equilateral hyperbola to calculate the logs of 1.1, 1.01, and a few more numbers to 55 places (*Mathematical Papers*, 1.101–41). Added to the procedures he inherited, the binomial expansion gave Newton a general method of quadratures by which he could find the area under virtually every algebraic curve then known to mathematics.

One of the central problems that the new analysis dealt with was drawing tangents to curves, essentially differentiation. In this case, Newton initially followed the approach of Descartes, which determined the normal to a curve, which is perpendicular to the tangent, at a given point. Always seeking to generalize, he worked out an algorithm valid for any equation to determine the subnormal, the distance along the x axis between the x co-ordinate of a point on a curve and the place where the normal to that point cuts the axis.

Patterns, which Newton did not fail to note, began to emerge. The formulae for the subnormals seemed related to those for squaring the same curves. The idea occurred: would it be possible to use the patterns to define curves that could be squared? As Newton pursued the idea he called upon the concept that areas under curves are generated by lines in motion. He was soon rewarded with what is recognized today as the fundamental theorem of the calculus—the inverse relation of differentiation and integration (*Mathematical Papers*, 1.221–33).

In the autumn of 1665 Newton extended the kinematic approach he had used for areas to tangents. He began to think of curves being generated by the motion of a point. If the equation of a curve expresses how x and y vary in relation to each other, the point that traces the curve will move at a rate compounded from the velocities of x and y, and the slope of the tangent at any point will be the ratio of their velocities. Let p be the velocity of x, and q the velocity of y. Newton derived an algorithm for determining the ratio of q/p (the derivative of y with respect to x) from the equation of the curve. His name for the method now beginning to emerge, the fluxional method (from the past participle of the Latin verb *fluere*, to flow) expressed the intuitive concept of motion with which he sought to escape from the complexities of infinitesimals, which he considered dubious concepts in mathematics.

The autumn of 1665 passed in feverish activity until, in a paper dated 13 November, Newton summarized his results, which solved the various problems in mathematics on which he had been focusing his attention. With their solution he seemed to lose interest and the manuscripts indicate that he did not touch mathematics again for six months. In May 1666 he returned to draft two successive versions of a general paper, and then, after another pause, a third version in October. He gave it the title 'To Resolve Problems by Motion these Following Propositions are Sufficient' (*Mathematical Papers*, 1.400–48). In the paper Newton spelt out the inverse algorithms for tangents and quadratures and applied them to a number of problems such as determining centres of curvature, and rectifying certain curves. Even though no one at the time knew of the existence of the tract of October 1666, Newton had become Europe's leading mathematician.

From the manuscripts it appears that Newton dropped mathematics once more and did not concern himself with it during the following two years. In the future he returned to mathematics at intervals: he worked on his fluxional method a number of times, mostly seeking to place it on a more secure foundation; he composed a path-breaking investigation of cubics; and he explored interpolation theory and projective geometry. As the years passed his returns to mathematics became less and less frequent, and increasingly it was external impulses that set him to work. As he said, he never again minded mathematics so closely.

Optics

There was other fare on which to feast. Whereas Cartesian and atomist philosophers considered light to be homogeneous, in the 'Quaestiones quaedam', as Newton considered colours, he had set down his first suggestion of the heterogeneity of light. He thought of an experiment to test whether the rays that provoke the sensation of blue may be refracted more than the rays that provoke red: colour one end of a thread red and the other blue, and when the thread is viewed through a prism the two ends should appear disjoined. He tried it, and they did. There, for the time being, he left the topic. By Newton's own testimony, somewhat later he was trying to grind lenses of elliptical and hyperbolic cross-section. Descartes had demonstrated that such lenses would refract parallel rays, those from a star for example, to a perfect focus, as spherical lenses do not. Newton realized that, while Descartes had assumed the homogeneity of light, the experiment with the thread seemed to demonstrate the opposite. Even if he succeeded in grinding the non-spherical lenses, they would not produce a perfect focus. At this point, Newton abandoned the lenses and took up an experimental investigation of the heterogeneity of light.

Newton's primary tool was an established instrument in optics, the prism. Into his darkened room he admitted a narrow beam of light through a hole in the shutter and refracted the beam through a prism onto a wall 22 feet away. This arrangement, especially the long trajectory giving enough distance for the refracted beam of finite cross-section to spread, altered the conditions of earlier experiments of which he had read and adapted the experiment to his own question. If light were homogeneous, the refracted beam ought to have cast a round image on the wall: instead it painted a spectrum five times as long as it was wide.

Newton recorded his investigation in an essay he entitled 'Of colours', entered into another of his student notebooks, probably early in 1666 (CUL, Add. MS 3975, 1–20). The new theory he began to elaborate inverted a tradition of more than 2000 years in Western science and philosophy, according to which light as it comes from the sun, pure and homogeneous, either is or appears white and colours arise from its modification. For Newton, light as it comes from the sun is a heterogeneous mixture of rays that differ in the sensations of colour they cause. Sensations of white and shades of grey are caused by the mixture; phenomena of colour arise from the separation of the mixture into its components, which for Newton are primary and immutable. Because the rays differ also in their degree of refrangibility, the prism disperses the incident beam into a diverging spectrum.

At that time Newton did not carry his work in optics to the same level of completeness that his mathematics had reached. The heterogeneity of light he subjected to a rigorous investigation, devising other experiments to separate white light into its components. He carefully eliminated the possibility that the spreading spectrum could be an accidental result of imperfections in the glass. Colours analogous to the prismatic spectrum, however, are a tiny fraction of all the phenomena of colours, most of which are associated with solid bodies. Newton was convinced that the other phenomena also arise from processes of separation, caused by the differing tendencies of rays to be reflected more readily from some surfaces than from others. A couple of experiments in the essay 'Of colours' spoke to this issue, one of them the first, relatively crude, production of Newton's rings, which would be, in a more refined state, the foundation of his approach to the colours of solid bodies. Another experiment, also relatively crude in comparison to later ones, illustrated the reassembly of the separated rays to produce the sensation of white—a demonstration essential to the theory. Not wholly free from the long tradition, Newton still spoke at this time largely in terms of two colours, red and blue. He had seized what remained the central idea of all his work in optics, the heterogeneity of light, but it is not wholly true that, in the language of his later statement, he 'had the Theory of Colours' in 1666.

Mechanics

During roughly the same period Newton also took his first steps in mechanics. Two of the 'Quaestiones quaedam' dealt with motion, and he had seemed at that point to embrace the atomistic doctrine that a principle of motion inheres in bodies. Not long thereafter, in another notebook, which he called the 'Waste book', he addressed the problem of impact (CUL, Add. MS 4004, fols. 10–15, 38*v*). Impact was a Cartesian problem. Newton considered it within a Cartesian context that started with the enunciation, in language almost identical to that in Descartes's *Principles of Philosophy*, of the principle that particles at rest or in motion retain their status unless disturbed by another body. Beyond this point, however, he diverged from Descartes, who had treated impact in terms of the force that a body in motion possesses by virtue of its motion. In contrast Newton began to define an abstract concept of force which, whatever its cause in a given case might be, set the change of motion proportional to the force exerted. He reasoned that when two bodies meet in impact each presses equally on the other, and both must undergo equal and opposite changes of motion. From this he concluded that the common centre of gravity of two isolated bodies in impact remains in an inertial state of rest or of uniform motion in a straight line. The conclusion is identical to the principle of the conservation of momentum. It appears as corollaries three and four to the laws of motion in his later masterpiece, the *Principia*.

On another problem in mechanics, circular motion, Newton remained closer to Descartes. The French philosopher, whose concept of today's 'rectilinear inertia' made circular motion a problem, set the pattern for its subsequent treatment with the idea that bodies constrained in circular motion endeavour constantly to recede from the centre. Centrifugal force, as Christiaan Huygens named it, was another force that a body possesses by virtue of its motion, but with circular motion Newton retained the concept that he replaced when he dealt with impact. Proceeding beyond Descartes he derived a quantitative measure of it, equivalent to today's formula for the radial force, mv^2/r.

Two other early manuscripts concerned with circular motion survive. In one of them Newton took up the problem he found in Galileo's *Dialogue*, an answer (imperfect in Galileo) to the argument that the earth cannot be turning daily on its axis because loose objects would be hurled into space. Newton saw that the objection could be assessed definitively with his quantitative measure of centrifugal force. From the accepted size of the earth he computed the centrifugal acceleration at the equator due to the diurnal rotation and compared it to the figure Galileo gave for the acceleration of gravity. Gravity, he found, is about 150 times greater than the centrifugal force. Then he saw that the same measure of centrifugal force gave him a means to check Galileo's figure for g, via the measured period of a conical pendulum. Galileo's value, it turned out, was only about half the correct one, and the objection against Copernican astronomy was doubly invalid.

In the other paper Newton substituted the values of Johannes Kepler's third law into his formula for centrifugal force and found that among the planets the force varies inversely as the square of their distance from the sun. He also compared the centrifugal force exerted by the moon in its orbit with the measured acceleration (or force for a given body) of gravity at the surface of the earth. The ratio came out as 1 to something more than 4000. In accordance with astronomical measurements, Newton was setting the distance of the moon at sixty terrestrial radii. An inverse square relation would have been 1/3600, and if a correct measurement of the earth had been available, he would have reached something close to this number. Most scholars believe that this paper was what Newton had in mind when he later stated that in the comparison he 'found them answer pretty nearly'. However, that does not mean that Newton had the concept of universal gravitation in 1666—nothing in the paper spoke of an attraction. He compared gravity at the surface of the earth with the moon's endeavour to recede, and if he silently assumed that something must be checking that endeavour since the moon remains in its orbit, nevertheless nothing like a universal attraction was mentioned and the correlation proved far from exact. Of his three early investigations, that in mechanics remained the least developed. Nevertheless, he remembered the calculation and later thought that it marked a step forward in his thought.

For all its brilliance the work in mathematics, optics, and mechanics which began during the final year or eighteen months of Newton's undergraduate career nearly blighted his future prospects. To pursue his new interests he had abandoned the established curriculum. In April 1664 Trinity had an election to scholarships, the only election during Newton's undergraduate years; only those who held scholarships would later be eligible for fellowships. Whatever obstacle his private course of study presented at this point, it appears that Newton had a patron in the college, probably Humphrey Babington, and he was elected. A little over a year later he proceeded to his BA degree.

Soon thereafter plague afflicted Cambridge and before the summer was over the university had dispersed. Newton returned to Woolsthorpe. The following March, after no deaths had been reported in Cambridge for six weeks, the university convened again, but by June it was evident that 1666 would also be a plague year. Only in the spring of 1667 did the university fully resume.

Much has been made of the plague years in the Newton legend. The *anni mirabiles* have been connected with his respite at home, but the manuscript evidence does not entirely support this notion. It is true that some of his important steps were taken in Woolsthorpe. The story of the apple—that, sitting in his orchard, he observed an apple fall from its tree and from this inferred the law of gravity—was set in Woolsthorpe and was connected with his idea that gravity extends to the sphere of the moon. Newton repeated the story to the antiquary William Stukeley in his old age and it was later retailed by his niece and her husband, John Conduitt, and by the natural philosopher Martin Folkes. Moreover, the tract of October 1666, his early definitive statement of the fluxional method, had to stem from Lincolnshire. Nevertheless, all of his investigations had begun in Cambridge, where the books that put him in contact with earlier figures of the scientific revolution were available, and his manuscripts do not reveal any marked break in continuity associated with the move. The marvellous years had as much to do with a young man's expanding intellectual world as with his physical location.

knew about his accomplishments, it led him to secure Newton's appointment to succeed him as Lucasian professor that same year, 1669. The income from the chair, about £100 per annum, added to his income from the Trinity fellowship and from his estate, ensured Newton's material well-being while he remained in Cambridge, and saved him, because he did not need the income which ordination would have provided, from having to subscribe to the articles of the Church of England, a requirement for fellows, but one which Newton would have been unable honestly to fulfil.

The statutes of the Lucasian chair required one lecture per week during three terms each year. Barrow had silently reduced the duty to one term per year, in which Newton acquiesced. Although the evidence is ambiguous, apparently he did deliver an annual series, often to an empty hall, until either 1684 or 1687. After that he converted the position into a sinecure, as other professors were doing, until he resigned it, together with his fellowship, five years after he left the university.

In 1669 Newton chose optics for his first lectures, and at that time he polished the theory of colours into its final form. The 'Lectiones opticae' ('Optical lectures'), deposited in the Cambridge University Library as the text of his first four sets of lectures, contains all the content of book one of the ultimate *Opticks*. About then he also returned to the experiment with Newton's rings, as they are still called, which would fill book two. He pressed a lens with a 50 foot radius of curvature down on a flat sheet of glass and with dividers carefully measured the diameters of the rings he observed in the film of air between the two pieces of glass. The computed thickness of the film marked the first secure entry of science into dimensions in the order of $^1/_{100,000}$ of an inch. The measurements also first established the periodicity of an optical phenomenon, although Newton did not hold that light itself is periodic. The periods belonged to waves of compression in the ether, generated in his view when a corpuscle of light strikes the first surface of a film, which determine whether the corpuscle can proceed through the second surface. Newton was convinced that the films, which reveal spectra, correspond to the thickness of the particles that compose bodies, and his experiments with the rings supplied his explanation of the colours of solid bodies.

The Royal Society and controversy

The renewed work on optics also brought Newton further onto the public stage. From the differing refrangibility of rays he concluded that chromatic aberration could not readily be eliminated from refracting telescopes. A reflecting telescope, however, would not have this defect, and Newton constructed the first known reflecting telescope early in 1669, and later a second. The telescope was about 6 inches long; it magnified about forty diameters. He was proud of his creation and showed it off, and by late 1671 news of it had reached the Royal Society in London, which asked to see the instrument. When it arrived at the end of the year it caused a sensation. The society sent notices about it to the scientific community on the continent and elected Newton a fellow on 11 January 1672.

Pleased by the response, Newton offered to send the society the discovery that had led him to the telescope—'in my Judgment the oddest if not the most considerable detection wch hath hitherto beene made in the operations of Nature'. The letter he sent on 6 February 1672 was a condensed statement of his theory of colours and some of the experiments on which it rested (*Correspondence*, 1.92–102), and the society published the paper in its *Philosophical Transactions*. His exposition effectively concealed the complicated steps by which he had arrived at his new and revolutionary theory, an approach which led to many favourable but some unfavourable responses.

Newton's equanimity quickly dissolved. Most important in this process was a condescending critique by Robert Hooke, considered the Royal Society's resident expert in optics. After nearly four months of mounting anger with Hooke, Newton sent a long reply couched in deliberately insulting terms. Through the society's secretary, Henry Oldenburg, he also received four successive letters in which Christiaan Huygens, the recognized doyen of European science, expressed his increasing reservations about Newton's theory. With Huygens, Newton did not allow himself to take the high tone he used with Hooke; nevertheless Huygens recognized Newton's irritation and refused to discuss the question further. There were also other letters (through Oldenburg) of query and criticism to which he had to reply. The number was not huge, but Newton was quickly complaining that he had sacrificed his peace, 'a matter of real substance' (*Correspondence*, 2.133). Only thirteen months after he sent this first paper to the Royal Society he requested that he be dropped from its rolls, and he informed Oldenburg that he intended 'to be no further sollicitous about matters of Philosophy' (ibid., 1.294–5). He did not pursue the resignation further, and Oldenburg simply allowed the matter to drop. Nevertheless, Newton did sever his correspondence with Oldenburg and Collins, who had been the channels for his communication with everyone else, and did attempt to regain his isolation. It was too late: the telescope and the letter on colours had introduced him to the larger world of scientific learning, which did not forget that a man of superlative quality lived and worked in Cambridge.

At the end of August 1674, when Newton was in London for a week, he made no effort to contact anyone connected with the Royal Society. When he was again in the city early the following year, however, he chose to attend the society. He found himself an object, not of criticism, but of respect. While he was there a letter arrived from Father Linus, an English Jesuit in Liège, denying that the experiment with the prism worked as Newton described it. Hooke informed the society that the experiment was beyond question, and the whole society made its support of Newton clear. He also met and conversed with Robert Boyle, whose works he had studied carefully. When Linus wrote again to the same effect in the autumn of 1675, Newton was sufficiently confident to offer another paper on optics. In fact, he sent two: a 'Discourse of observations', on the colours of thin films, which was virtually identical to the first three parts of book two of the eventual *Opticks*, and 'An hypothesis explaining the properties of light'. Based on the concept of a universal ether, the 'Hypothesis' did not confine itself to optical phenomena. It presented a general system of nature, beholden to mechanical philosophies but also pervaded with themes that derived from his recent study of alchemy, which presaged future developments in Newton's natural philosophy.

The 'Hypothesis' quickly involved Newton in new controversy. It contained two provocative references to Hooke, and at the conclusion of its reading before the Royal Society in December 1675 Hooke rose to assert that most of its content was contained in his own *Micrographia*. When the charge was relayed to Cambridge, Newton exploded in rage reminiscent of that in 1672.

A letter from Hooke, who claimed that Oldenburg had misrepresented him and went on to express his esteem for Newton, led to a mollified response in which Newton likewise expressed his esteem. Their basic antagonism remained, however; it would later find occasion for further expression.

Meanwhile the controversy with English Jesuits in Liège, arising from his paper of 1672, continued to disturb Newton's peace. When Linus died his student Anthony Lucas took up the cause and soon drove Newton to distraction. 'I see I have made my self a slave to Philosophy', he stormed to Oldenburg, after he had written only a fifth letter, 'but if I get free of Mr Linus's business I will resolutely bid adew to it eternally, excepting what I do for my privat satisfaction or leave to come out after me' (*Correspondence*, 2.182–3). Eventually, after a brutal letter in which he poured his paranoia over Lucas, he refused to receive any further communication from Liège and once more attempted to isolate himself from the learned world.

It was during this period of stress that Newton learned of another man who would later figure prominently in his life, Gottfried Wilhelm

Leibniz, and briefly corresponded with him, as usual not directly but through an intermediary. In May 1676 in a letter to Oldenburg, Leibniz, who had developed his system of calculus the previous autumn, asked for a derivation of two infinite series by Newton that he had received from Collins. Urged by Oldenburg and Collins, Newton composed a long letter on infinite series. Later that year, in October, he composed a second letter in response to Leibniz's enthusiastic reception of the first (*Correspondence*, 2.20–32, 110–61). Nearly forty years later, when the priority dispute raged, Newton called upon the two letters, which he referred to at that time as 'Epistola prior' and 'Epistola posterior', as evidence against Leibniz. In the 'Epistola posterior', Newton indeed approached the fluxional method, brief statements of which he concealed in indecipherable anagrams. By the time of the controversy, Newton had learned further that, on a visit to London, Leibniz also saw papers by him in Collins's possession. Not only did Newton not meet the German mathematician at the time, he rebuffed Leibniz's eager desire to communicate. Prone in any case to isolation and thoroughly agitated then both by Hooke and by Lucas, he never replied to Leibniz's response to his second epistle.

Instead of an intellectual exchange at the time, he earned a bitter dispute forty years later, when he used the 'Epistolae' as ammunition against his opponent.

Alchemy and theology

When Newton told Oldenburg that he intended to bid philosophy adieu, he was not issuing an idle threat; well before 1676 new interests had seized his attention. One of them was alchemy. Shortly after he composed the 'Quaestiones quaedam', Newton began to read earnestly in chemistry, and about 1669 the interest turned decidedly toward alchemy. On a trip to London he purchased the massive six-volume *Theatrum chemicum* of 1602, and in the months ahead pored over its collection of alchemical treatises, taking extensive notes. Both here and elsewhere he studied all the recognized authorities, including contemporaries such as the pseudonymous Eirenaeus Philalethes (George Starkey), one of Newton's favourite authors. One student of alchemy asserts that the vast literature of the art has never been read so thoroughly. Newton assembled one of the great collections of alchemical books, which formed a significant proportion of his personal library; he made contact with clandestine circles of alchemists from whom he received manuscripts that he copied; and in the garden outside his chamber in Trinity he set up a laboratory in which he experimented.

Some time about 1670 Newton compiled a manuscript known as 'The vegetation of metals' from a phrase in its first line—'that metalls vegetate after the same laws [of Nature]'. In the paper, which is more a collection of thoughts than a connected discourse, he insisted on the distinction between mere mechanical alterations in the textures of bodies and the more subtle and noble changes wrought by vegetation. The principles of vegetable actions are 'the seeds or seminall virtues of things those are her only agents, her fire, her soule, her life' (Smithsonian Institution, Dibner Collection, Burndy MS 16). Some of the passages in 'Vegetation' are similar to parts of the 'Hypothesis ... of light', composed not long after it and pervaded with themes from alchemy: in the latter Newton met a natural philosophy that spoke in terms of life and spirit rather than inert particles in motion, and in the two manuscripts its influence began to transform his mechanical philosophy of nature.

Alchemy was not Newton's only new study—there was also theology. The statutes of Trinity College may have started his serious reading in theology. They required that all fellows, with the exception of the holders of two specified fellowships, be ordained to the Church of England ministry within seven years of incepting MA. Newton was never one to take an obligation lightly, and it may have been the approaching ordination that set him on serious study of theology, about 1670. In his usual style he purchased a notebook and entered a set of headings under which to collect the fruits of his reading in an orderly way (King's Cam., Keynes MS 2). He devoured the Bible, making himself a master of it to an extent that few could match, and tackled the early fathers of the church in a prodigious programme of reading that took him through all the major fathers and many lesser ones as well. Almost immediately his study found a focus. In his notebook headings such as 'Christi passio, descensus, et resurrectio' and 'Christi satisfactio, & redemptio', apparently expected to be major topics from the space allotted to them in anticipation, received very few entries. 'Deus filius' ('God the Son'), on the other hand, spilled over the smaller space originally intended for it, and the entries he did set down suggest that very early he began to see a distinction between God the Father and God the Son and to question the status of Christ and the doctrine of the Trinity.

It did not take Newton long to read himself right out of orthodoxy. He became fascinated with the theological struggle of the fourth century as a result of which trinitarianism was established as Christian orthodoxy. For Athanasius, the principal architect of trinitarianism, he developed more than a mere antipathy—passionate hatred is a better description. One of his manuscripts, 'Paradoxical questions concerning the morals & actions of Athanasius & his followers' (King's Cam., Keynes MS 10), virtually stood Athanasius in the dock and prosecuted him for an extended litany of sins. Newton enlisted himself among the disciples of Athanasius's opponent, Arius, for whom Christ was not an eternal part of the Godhead but a created intermediary between God and man, a doctrine similar but not identical to modern unitarianism.

Interpretation of the prophecies formed a major strand of Newton's theological quest—primarily the book of Revelation at first, but later Daniel as well. Newton's interest in the prophecies is well known from a volume published after his death, but that publication was a product of his old age, sanitized by him to obscure its point. The unpublished interpretation that he composed in the 1670s fitted harmoniously with his new theological stance. The prophecies foretold the great apostasy, the rise of trinitarianism. The vials of wrath of the apocalypse represent the barbarian invasions of the Roman empire—'like Furies sent in by the wrath of God to scourge ye Romans'—God's punishments of a stiff-necked people who had gone whoring after false gods (Jewish National and University Library, Jerusalem, Yahuda MS 1.4, fol. 127).

Heterodoxy and heresy

Newton's Arianism nearly terminated his academic career. Religious heterodoxy of that extent was not something Cambridge would tolerate. Not that he advertised his heretical views—far from it. Newton understood perfectly that his theological views were anathema to the established order, and the need to keep what he regarded as his most important truth to himself was a central ingredient in his notorious secretiveness. The problem was ordination, required by 1675 at the latest:

Newton had moved himself beyond the possibility of ordination and in the questions that would have arisen Newton could not have concealed his heretical views. A potential avenue of escape presented itself when one of the two exempted fellowships fell vacant, but unfortunately a fellow more senior than Newton claimed it. In a letter to Oldenburg early in 1675 Newton indicated that he was laying down his fellowship. At the last minute, probably through the intervention of Isaac Barrow, by then master of Trinity, a royal mandate that exempted the Lucasian professor in perpetuity from any college requirement of ordination rescued Newton and enabled him to continue in his sanctuary, at the time devoting himself largely to writing Arian tracts and to pursuing alchemy, two activities that he indulged in private.

In the early 1680s Newton went beyond mere Arianism in his most important theological composition, 'Theologiae gentilis origines philosophicae' ('The philosophical origins of gentile theology'), as he called it in the least chaotic of its incomplete manuscripts (Jewish National and University Library, Yahuda MS 16.2). The 'Origines' removed the coming of Christ from the focus of world history and treated him as merely the latest in a series of prophets sent by God to reclaim mankind from false gods. But mankind has an innate tendency to idolatry; trinitarianism, the worship of a creature as God, was only another turn in the cycle that throughout history had repeatedly perverted worship. 'What was the true religion of the sons of Noah before it began to be corrupted by the worship of false Gods', Newton wrote as the title to one chapter that he did not compose; 'And that the Christian religion was not more true and did not become less corrupt' (Yahuda MS 16.2, fol. 45*v*). In the years ahead Newton kept inserting perplexing passages drawn from the 'Origines' into his scientific work—the final paragraph of 'Query 31' in the *Opticks*, for example, and the two footnotes to the 'General scholium' added to the *Principia*. Thoroughly purged, it became his *Chronology of Ancient Kingdoms Amended* (1728). The original treatise he kept to himself.

Alchemical work continued in tandem with theology. In the early 1680s Newton began what he called the 'Index chemicus', a typical Newtonian exercise in organizing and systematizing his knowledge. Newton kept expanding the 'Index' for more than a decade. In its ultimate form—there is nothing to suggest that he thought it complete—it contained 879 separate entries that cited more than 150 separate alchemical works, extending from mythical alchemists of antiquity to Newton's contemporaries, and gave about 5000 separate page references to them (King's Cam., Keynes MS 30). He continued also to experiment. His dated experimental notes began in 1678 and continued, with a break for composition of the *Principia*, almost until his departure from Cambridge (CUL, Add. MSS 3973; 3975, 101–58, 267–83). Couched in Newton's personal set of symbols, the experimental notes have so far resisted interpretation. Now and then, however, when success seemed to crown his endeavours, he interjected notes of exultation that echo the imagery of alchemy: 'Friday May 23 [1684] I made Jupiter fly on his eagle' (CUL, Add. MS 3975, 149).

Alchemy led to an important correspondence with Robert Boyle, his first significant direct correspondence with no intermediary. In a long letter to Boyle that Newton composed in 1679 he argued, among other things, for 'a certain secret principle in nature by wch liquors are sociable to some things & unsociable to others' (*Correspondence*,

2.288–95). A paper, 'De aere et aethere', related to the letter to Boyle, asserted that particles of air repel each other 'with a certain large force' (*Unpublished Scientific Papers*, 214–20). Mechanical philosophers employed an invisible ether to explain such phenomena. The letter to Boyle introduced an ether, though it also denied that the secret principle of sociability had anything to do with the sizes of particles and pores. The second chapter of 'De aere et aethere' also started to describe such an ether. After a few lines, however, it stopped in the middle of a sentence, as though he found the exercise futile. During the following thirty-five years the concept of an ether disappeared from Newton's natural philosophy, which dealt now with forces between particles.

About this time Newton underwent a general revulsion from Descartes, the philosopher who, more than anyone else, had introduced him to the new world of scientific thought, including the mechanical philosophy. The revulsion included mathematics: Cartesian geometry, he now declared, was 'the Analysis of the Bunglers in Mathematicks' (Hiscock, 42). Natural philosophy and religion were central to the revulsion. An essay, 'De gravitatione et equipondio

48 fluidorum' ('On the gravity and equilibrium of fluids'), not only repudiated basic themes of Cartesian philosophy, with a passion reminiscent of his attacks on Athanasius, but accused Descartes of atheism as well (*Unpublished Scientific Papers*, 89–121).

Principia mathematica

4

Planets and comets

Newton's mother died during the spring of 1679 and he was absent from Cambridge for six months, tending her during her final days and then settling the estate in Woolsthorpe. Immediately upon his return late in November a letter from Robert Hooke arrived, the first of a number of intrusions prompted by his earlier renown in mathematics and natural philosophy. Hooke, now secretary of the Royal Society in place of the recently deceased Henry Oldenburg, invited Newton to resume correspondence with the society. Towards that end he invited Newton's opinion on his theory that planetary motions are compounded of a tangential motion that is continually deflected by an attraction towards a central body, the remarkable idea that Hooke had expounded in his Cutlerian lecture of 1674. It proposed the idea of a general attraction, not quite universal gravitation, and it inverted the conceptualization of circular motion by focusing attention, not on centrifugal force, but on an attraction towards the centre. Every indication is that Hooke, at this point, taught Newton to conceive of circular motion in these new terms. Not long thereafter Newton coined the phrase 'centripetal force', in imitation of Huygens's word 'centrifugal'; the central theme of the *Principia* would be the quantitative exploration of the new concept, something Hooke had not achieved.

At the time Newton rejected the correspondence; he was engaged in other studies, he told Hooke, and grudged time spent on philosophy. Nevertheless, he went on to propose an experiment to demonstrate the rotation of the earth. Opponents of Copernicanism argued that bodies dropped on a turning earth would appear to fall to the west; Newton proposed that a body dropped from a high place would fall rather a tiny distance to the east because the top of a tower has a higher tangential velocity than its foot, and treating the earth as a non-resisting vacuum, he drew the path as part of a spiral that ended at the centre.

That is, he converted the problem of fall on a rotating earth into the problem of orbital motion conceived according to Hooke's idea, and in the process he made a monumental blunder. Hooke corrected him; under conditions of no resistance, the body dropped would not tend to the centre but would rather return to its original height, and the path would be a sort of ellipse. In turn, Newton corrected Hooke; given a constant attraction, the body would return to its original height, not at 360°, but at about 240°. Hooke replied once more that he did not assume a constant gravity, but a force that varies inversely as the square of the distance. Here the correspondence ended. Seven years later it formed the basis of Hooke's charge of plagiarism, not of the dynamics of circular motion, but of the inverse square relation. At the time it led Newton to demonstrate for his own satisfaction that an elliptical orbit with the centre of attraction at one focus entails an inverse square force.

A year later it was nature which intruded, by way of two comets, the first one visible before dawn, then, after an interval, another in the evening sky. No one had seen a comet greater than the evening one; it stretched some 70 degrees across the sky. Within days of its first sighting Newton began to observe it closely, and systematically collected data both on this comet and on others. Through an intermediary he also corresponded about it with John Flamsteed, the astronomer royal, who was convinced that the two appearances were not two comets but a single one which reversed its direction in the vicinity of the sun. He expounded the theory in terms of a fantastic magnetic dynamics, rejected by Newton, who also resisted the notion of a single comet. Although he had recently worked out the dynamics of orbital motion, he did not apply it to the comet at this time. Comets had always been regarded as foreign bodies, altogether different from planets. Whatever ideas of attractions Newton was entertaining in 1681, they did not add up to universal gravitation.

In August 1684 Newton received a visit from Edmond Halley, and this intrusion proved to be decisive. In London, Halley had been in a discussion with Christopher Wren and Hooke about the shape of an orbit in an inverse square force field. Clearly Hooke was misled in thinking that he alone had come upon the inverse square relation. Newton had derived it twenty years earlier, and apparently Halley and Wren had done the same more recently. Both Halley and Wren admitted that they could not work out the orbital dynamics, and although Hooke claimed that he could, he did not produce a demonstration. Halley, in Cambridge, put the question to Newton, who replied at once that he had demonstrated that an orbit in an inverse square attraction must be an ellipse. When he failed to find the paper, he promised to work out the demonstration anew and send it to Halley.

As a consequence, in November, Halley received a nine-page tract known as 'De motu' ('Concerning motion'), which sketched in an orbital dynamics virtually identical to that which appears in the *Principia*. Already Newton was entertaining ideas that stretched beyond orbital dynamics. He asked Flamsteed for information about the satellites of Jupiter and Saturn and about the motion of Saturn as Jupiter approaches conjunction with it. In fact the issue of celestial motion had seized Newton's attention and expelled alchemy and theology entirely. For the following two and a half years Newton virtually cut himself off from society to pursue a problem which kept expanding in every direction and revealing new facets. Swept along by the grandeur of the theme, he finally saw a work to completion. It transformed his life, and it transformed science.

Motion and gravitation

The first issue was dynamics. 'De motu' sketched an orbital dynamics but not a system of dynamics. During the early months of 1685 Newton gradually worked out what became his three laws of motion. For twenty years he had hesitated before the principle of inertia he had found in Descartes, now seeming to embrace it, now retreating, especially as he recoiled from Cartesian relativism, to the notion that a force internal to bodies sustains their motion. As he pondered over dynamics early in 1685 he realized once and for all that he could not elaborate a consistent dynamics without the principle of inertia. His first law of motion remains today the classic statement of the principle: 'Every body continues in its state of rest, or maintains motion in a right line, unless it is compelled to change that state by forces impressed upon it.' For the second law he returned to the concept he had begun to explore in 1665 as he treated impact: the change of motion is proportional to the force impressed. The law implied a concept of quantity of matter, which Newton defined at this time and named 'mass'. The third law generalized the equal and opposite changes of motion in impact. Before the middle of 1685 Newton had defined a system of dynamics from which both Galileo's kinematics of terrestrial motion and Kepler's kinematics of celestial motion emerged as necessary consequences.

And he arrived at the concept of universal gravitation. Obviously Newton's thoughts were moving in that direction when he asked Flamsteed about the motions of Jupiter and Saturn in conjunction and when he began, as he did in 'De motu', to treat comets as planet-like bodies governed by the same attraction. Details made it more than a general idea. He realized that with pendulums he could, with great precision, test Galileo's assertion that all bodies fall with the same acceleration. 'When experiments were carefully made with gold, silver, lead, glass, sand, common salt, water, wood, and wheat', his revisions of 'De motu' record, pendulums of identical length had identical periods (Herivel, 311, 319). This is only possible if the earth attracts all the particles in those various substances in exact proportion to their quantities of matter. Kepler's third law implies that the sun attracts the planets and Jupiter attracts its satellites in exact proportion to their masses, and if the satellites remain concentric with Jupiter, the sun must attract both them and Jupiter in exact proportion to their masses. Because the moon stays in orbit around the earth, the earth must attract it, and in 1685, with a correct measure of the earth, the correlation of the moon's orbit with the measured acceleration of gravity on the earth was not pretty near but exact within an inch in the measure of *g*. By the third law of motion, all these orbiting bodies must also attract the bodies they orbit. Some time early in 1685 Newton concluded that every particle of matter in the universe attracts every other with a force proportional to the product of their masses and inversely proportional to the square of the distance between them.

Writing *Principia*

Clearly an idea of this magnitude demanded more than a nine-page tract, and during the following two years Newton expanded 'De motu', first into an intermediate treatise in two books, and then into the three-book treatise we know. He confronted the problem inherent in his correlation of the moon's orbit with *g*, that he was treating the apple (in the well-known story) as though it were attracted not to the surface but to the centre of the earth. In one of the key propositions of the *Principia* Newton demonstrated that the composite attraction exerted on external bodies by a homogeneous sphere, made up of particles that attract with a force inversely proportional to the square of the distance, is directed toward its centre and varies according to the same law. He developed an analysis of the three-body problem that allowed him to argue with justice that most of the perturbations of Kepler's laws arising from the mutual attractions of bodies in the solar system fell below the threshold of observation as it then stood. By April 1686 he was sufficiently satisfied with book one to send the manuscript to the Royal Society.

In response Halley assured him that his 'Incomparable treatise' had arrived and that the society greatly appreciated the honour of being its dedicatee. Unfortunately, he had also to report that Hooke had raised the cry of plagiarism. Initially Newton reacted calmly enough. After he had fed on the accusation for three weeks, however, he was in a flaming rage:

> Now is this not very fine? Mathematicians that find out, settle & do all the business must content themselves with being nothing but dry calculators & drudges & another that does nothing but pretend & grasp at all things must carry away all the invention. (*Correspondence*, 2.435–40)

He then informed Halley that 'Philosophy is such an impertinently litigious Lady that a man had as good be engaged in Law suits as have to do with her' (ibid.). One of his latest revisions of book three had been a handsome acknowledgement of Hooke's concept of attraction. He now slashed it out and removed every other reference to Hooke in the manuscript except for a few indispensable observations.

Newton also threatened to suppress book three, a matter of consequence for Halley, who was in fact the publisher after the nearly bankrupt Royal Society had handed the manuscript to him. What Newton now threatened to suppress would have mutilated the work to which Halley was devoting not only his energy but his meagre resources. By diplomatic suasion Halley calmed the storm, but Newton did not withdraw his threat. It was only in the spring of 1687, after he had received the copy for book two, that Halley learned what the full content of the book he was publishing would be.

'De motu' had contained two propositions about motion through resisting media. Newton greatly expanded his treatment of this problem, to which he finally devoted book two, and altered its thrust into an attack on Cartesian natural philosophy. On the one hand new propositions on the resistance of physical media showed how quickly a dense medium, like that which filled the Cartesian universe, would bring planets to rest. On the other hand an analysis of vortical motion demonstrated that vortices are incompatible with Kepler's laws and that they cannot be self-sustaining. The hypothesis of vortices, he concluded, 'is utterly irreconcilable with astronomical phenomena and rather serves to perplex than explain the heavenly motions' (*Mathematical Principles*, 396).

Book three proceeded then to explain the heavenly motions in terms of attractions propagated through empty space. The early part of the book derived the law of universal gravitation in much the way presented above. Newton then proceeded to employ the concept to explain, in quantitative terms, a number of phenomena not employed in its derivation. He redirected the analysis of the three-body problem, developed originally to argue that most perturbations are below the threshold of observation, to submit some perturbations to exact calculation. Astronomers had long known of a number of irregularities in the moon's motion. By demonstrating how the external attraction of the sun would affect the moon's orbit around the earth, Newton shifted lunar theory from empirical description to dynamic analysis. A similar analysis, in which the moon and sun together were the external sources of perturbation and the water of the oceans was the body around the earth, offered the first satisfactory explanation of tides. Applied to the bulge of matter around the equator, the analysis derived a conical motion of the earth's axis that gives rise to an appearance known as precession of the equinoxes. In a final *tour de force*, in what Newton called the most difficult demonstration in the work, he succeeded in reducing the observed positions of the great comet of 1680–81 to a conical orbit, a parabola in the first edition, later an ellipse.

At a number of places in the *Principia* Newton inserted disclaimers that his book intended to assert the reality of attractions; it merely pursued the mathematical description of observed motions. These assertions were a vain attempt to forestall objections by mechanical philosophers to the very concept of attraction at a distance. In fact, Newton was reshaping natural philosophy into a new mould that allowed mathematical treatment in a way that etherial mechanisms never could. As he drew his book to a conclusion, he considered adding a general discussion of forces. The *Principia* dealt with the great motions readily observed in the heavens:

> There are however innumerable other local motions which on account of the minuteness of the moving particles cannot be detected, such as the motions of the particles in hot bodies, in fermenting bodies, in putrescent bodies, in growing bodies, in the organs of sensation and so forth. If any one shall have the good fortune to discover all these, I might almost say that he will have laid bare the whole nature of bodies so far as the mechanical causes of things are concerned. (*Unpublished Scientific Papers*, 333)

In the end he sought, also in vain, to avoid controversy by suppressing the discussion. In an expanded version it appeared twenty years later as 'Query 31' of the *Opticks*.

Halley received the manuscript for book two early in March 1687, and that for book three early in April. He engaged a second press and for four months did little except supervise the edition. On 5 July he was able to inform Newton that his *Philosophiae naturalis principia mathematica* ('Mathematical principles of natural philosophy') had appeared. Never in the history of civilization has a major theory been so fully, so clearly, or so influentially proved.

Revolution and breakdown

In February 1685, not long after Newton seriously immersed himself in the *Principia*, the death of Charles II passed the crown to his brother James II, a dedicated Roman Catholic committed to the reconversion of England to Roman Catholicism. The time it took for the crisis implicit in James's succession to reach Cambridge was precisely the time Newton needed to complete his work. On 9 February 1687, as he was polishing its final pages, a letter mandate arrived from James ordering the university to admit Alban Francis, a

Benedictine monk, to the degree of master of arts and thus to the senate of the university. The letter presaged a campaign to install a Roman Catholic majority in the university. Free from the burden his master-work imposed, Newton, that reclusive fellow of Trinity College, who had sought primarily to isolate himself from the university community but who also hated popery with passion, suddenly emerged as one of its champions in resistance to James.

In April, Newton was one of eight representatives of Cambridge called to answer in London before the notorious Judge Jeffreys, and two years later, after the uprising of 1688 had ratified his courage, he was elected as one of the two Cambridge representatives in the convention parliament. Without taking a leading role, Newton stood four square with the majority who declared that the Roman Catholic James had forfeited the throne, and in general he supported the principles the uprising installed in the English constitution. After the parliament was dissolved in 1690 Newton did not stand for re-election, though later, from December 1701 to May 1702, when he was a public servant, he represented Cambridge in parliament, and once more he was inconspicuous.

He stood a last time in 1705, only to be rejected in a tumult about conformity to the established church. Nothing could induce him thereafter to repeat that experience. Whatever his role in parliament, Newton's leadership in the resistance to King James in Cambridge, capping the triumph of the *Principia*, which had made him one of the recognized intellectual leaders of England, promoted the recluse of Trinity into one of the university's most prominent figures.

In London, moreover, the horizons of the recluse expanded. He made new friends. As a refugee in the Netherlands, the philosopher John Locke had grasped the significance of the *Principia* and when the 1688 action made it possible for him to return home he immediately sought out Newton's acquaintance. The two men had much in common—intellectual distinction of course, and a shared outlook on politics and on religion. Locke was probably the first man with whom Newton openly discussed his theological views, and for Locke had composed an essay, 'Two notable corruptions of scripture', in which he demonstrated that the two primary scriptural supports of trinitarianism had not appeared in the Bible before the fourth century. As he had done with

Boyle, Newton corresponded with Locke directly rather than through an intermediary, and they remained in contact until Locke's death in 1704.

In London, Newton also met a brilliant young Swiss mathematician, Nicolas Fatio de Duillier (1664–1753), who had recently arrived from the continent. Instantly the two became very close. Fatio arrived a Cartesian. Soon he was singing in a Newtonian key. For a time he entertained plans to edit a second edition of the *Principia*. He was one of the first men freely to explore Newton's mathematical papers, an exploration that led him to begin suggesting that Newton both preceded and excelled Leibniz in the calculus. Newton introduced Fatio to heterodox theology, the prophecies, and alchemy. The relationship, incandescent in its emotional intensity, dominated Newton's life for about five years. Fatio, who lived in London, broached the idea that Newton should find a governmental appointment there. In turn, when Fatio became ill late in 1692, Newton urged him to live with him in Cambridge, where he would provide an allowance.

There were other new friends. In the aftermath of the *Principia* young men who recognized its power sought to enlist themselves as Newton's epigoni. The Scottish mathematician and astronomer David Gregory gained Newton's support, which may have been critical, for appointment as Savilian professor of astronomy at Oxford. An aspiring young divine, Richard Bentley, who had been selected to deliver the first series of Boyle lectures in defence of religion, sought Newton's comments as he prepared for publication texts that drew heavily on Newtonian philosophy. In the four letters to Bentley that Newton eventually composed, he explored the deity's relation to the creation (*Correspondence*, 3.233–54). Somewhat later Newton came to know William Whiston, who would succeed him as Lucasian professor. Between 1707 and 1710, Whiston's public avowal of theological positions he probably learned from Newton led to his expulsion from Cambridge and threatened Newton with the exposure of matters he strove always to keep secret.

The early 1690s were a period of manic intellectual activity as Newton, realizing the *Principia*'s success, sought to produce a synthesis of the new natural philosophy in the Newtonian style. He began a second edition of the *Principia*. He drew his work in optics and in the calculus together. In the end, he abandoned all of these projects, at least for the time being. Theology was not prominent in his endeavour at that time; although he did compose 'Two notable corruptions' during those years, the *Principia* marked the beginning of a hiatus of about two decades in serious theological study. For alchemy, in contrast, the *Principia* was a momentary interruption, and Newton took up the art anew even before the *Principia* was completed. Roughly half of the large volume of alchemical manuscripts that he left behind dated from the years immediately after the *Principia.* He devoted time to expanding his 'Index chemicus'; he experimented; and he composed his most important alchemical paper, 'Praxis', which went through four successive versions and seemed to claim success in the alchemical work:

> Thus you may multiply each stone 4 times & no more [he wrote in the essay's climax] for they will then become oyles shining in ye dark & fit for magicall uses.... This is ye multiplication in quality. You may multiply it in quantity by the mercuries of wch you made it at first... Every multiplication will encreas it's vertue ten times &, if you use ye mercury of

> ye 2^{d} or 3^{d} reaction wth out ye spirit, perhaps a thousand times. Thus you may multiply to infinity. (MIT Cam., Dibner Institute, Babson MS 420, 18a)

When Newton composed the final version of 'Praxis' in the summer of 1693, he was in a state of acute tension. Beyond this text, some dated alchemical experiments, and a letter he began but did not finish, nothing is known of his activity that summer. In the middle of September he broke silence with two wild letters. He informed Samuel Pepys that he must withdraw from his acquaintance and see none of his other friends any more. To Locke he wrote that he was so much affected by the conviction that Locke had tried to embroil him with women 'that when one told me you were sickly & would not live I answered twere better if you were dead' (*Correspondence*, 3.279–80). Before long stories were circulating on the continent that Newton had suffered a mental derangement that lasted a number of months. Undoubtedly the stories were exaggerated, but just as undoubtedly there was a breakdown—difficult to define—during 1693. It brought an end to the euphoria of the early 1690s and to the plans for publication, and it effectively terminated

Newton's intellectual creativity. It may have terminated his involvement with alchemy; although he did date some experiments after 1693, he abandoned the work about the time he moved to London. It also marked the end of his close relationship with Fatio, an event shattering for Newton but even more so for Fatio, who soon abandoned science and mathematics and drifted into religious fanaticism.

In the aftermath of the breakdown Newton resolved to have, as he put it, another go at the moon. The detailed lunar theory had been one of the last additions to the *Principia*, where it was clearly imperfect. Newton intended to carry the investigation to completion, and now was the time to do so. To this end he applied to Flamsteed for observations of the moon, and in the summer of 1694 he set about the task in his usual intense way. It was excruciating work, dealing with a set of corrections that could hardly be distinguished from each other. He later told one associate that 'his head never ached but with his studies on the moon' (King's Cam., Keynes MS, 130.6, book 3). Measured by his aspiration to devise a theory accurate to 2 minutes, his efforts were a failure. Increasingly frustrated, he projected his failure onto Flamsteed, who had not, Newton claimed, supplied the observations he needed. Finally, a furious letter in July 1695 effectively ended amicable relations between the two, and not long thereafter Newton abandoned his effort.

The failure over the moon confirmed the breakdown of 1693. Realizing that his creativity was exhausted, Newton began to find the leisure Cambridge supplied more of a burden than a benefit, and when the opportunity of an appointment in London opened he seized it without hesitation. Within a month of receiving the proffered appointment, he had moved out, lock, stock, and barrel.

Newton in London

5

The Royal Mint

Newton had been consulted on matters connected with the Royal Mint in 1695. Through the combined efforts of clippers (who clipped silver from the edges of hammered coins that had no definite outline) and coiners (or counterfeiters), English money—the silver money of daily commerce—had deteriorated alarmingly in the early 1690s. The crisis in the coinage was made worse by the simultaneous financial crisis caused by the unprecedented expenses of a war with France. In an age when professional economists did not exist the government, in 1695, sought advice from eight men known for their intellectual prowess or their experience in matters financial. The fact that Newton was one of the eight testifies to the impact of the *Principia*. Together with another six he favoured recoinage. By December the government had made its decision, and the recoinage was already under way when, late in March 1696, Newton received a letter from Charles Montagu, chancellor of the exchequer and a prominent figure in the whig junto then in control of the government, offering him the wardenship of the mint. He did not hesitate in accepting and by 2 May had installed himself in London and assumed his duties.

Montagu's letter was unambiguous in offering a sinecure. Newton had never learned how to do things halfway, however, and his personal need to escape from intellectual activity become fruitless made common cause with the need of an institution in chaos under the demands that the recoinage imposed. He gave of himself without stint. Arranging and ordering were among Newton's most basic instincts. Every time he had approached a new discipline, his first step had been an ordered collection of information. The same impulse served him well at the mint. He was by nature an administrator, and as far as can be untangled from the confused record, it appears that his presence was a significant factor in bringing about an orderly functioning of the mint and radically increasing the rate of coinage until it reached £100,000 per week at a time when, with the old coins called in, the shortage of legal tender was threatening to bring economic life to a standstill. He appears certainly to have been instrumental in setting up the five country mints called for by parliament. When the recoinage ended in the summer of 1698 the mint had coined a total of £6.8 million in two and a half years, nearly twice the total coinage of the previous thirty. Some significant portion of that record was Newton's work.

Newton soon learned that there was another dimension to his duties as warden that he had not counted on. The warden was responsible for apprehending and prosecuting clippers (who soon ceased to exist with the new coinage) and coiners (as yesterday's clippers soon became). Newton disliked the job and initially petitioned to be relieved of it. When the Treasury would not even hear of that, he plunged in with his normal intensity: he had himself commissioned as a justice of the peace in all the home counties; he took innumerable depositions in taverns and in gaols from assorted riff-raff; and he successfully prosecuted twenty-eight coiners. In all more than a hundred were pursued under his authority.

As Newton accustomed himself to a new life, aspects of the old one continued to make their appearance. In 1697 he received two challenge problems from Johann Bernoulli, who intended to show that Newton's pretended, but unpublished, method was not as powerful as the differential calculus of Leibniz and Bernoulli. John Conduitt, the husband of Newton's niece, later heard the following story, which surviving documents support: Newton came home from the mint one evening, tired, to find the problems and solved them before he went to bed. When Bernoulli shortly thereafter received from England an anonymous paper with the solutions, he understood at once whence it came—'as the lion is recognized from his claw', in his classic phrase. Newton also chose to have another try at the moon. It got no further than the previous one, and led just as quickly to another angry confrontation with John Flamsteed. In 1698 there was an episode of a different sort; the French Académie des Sciences elected Newton as one of eight foreign associates.

London life

Initially Newton moved into the warden's house in the mint, which was located between the walls of the Tower of London. He did not stay there long. By August he had installed himself in a house on Jermyn Street in Westminster, where he lived for more than a decade. In 1709 he moved to Chelsea, quickly found it unsatisfactory, and the following year transferred to St Martin's Street, south of Leicester Fields (now Leicester Square). Here he stayed until, with his health decaying in his final years, he moved to Kensington. If he did not live sumptuously, he certainly lived well, keeping a coach until his last years and a fleet of servants.

Soon after he settled in London, although the exact date is not known, Newton's niece Catherine Barton went to live with him. She was thereafter a constant part of his life, from 1717 with John Conduitt, her husband, and their daughter. Catherine Barton was a young woman approaching twenty when she arrived in London, beautiful, lively, witty, and possessed of unlimited charm. Among those who felt the charm was Charles Montagu, the earl of Halifax as he became, Newton's old friend who had secured his appointment at the mint. The relationship of Catherine Barton and Halifax was known to the gossips of London, who passed it on to Voltaire, who later broadcast it to the world. It was not merely gossip. In 1706 Halifax drew up a codicil to his will leaving Catherine £3000 and all his jewels 'as a small Token of the great Love and Affection I have long had for her' (Earl of Halifax [C. Montagu], *Works... Earl of Halifax*, 1715, 1716, iv–vi). Seven years later a second codicil replaced the first, now bequeathing £5000 plus a grant during her life of the rangership and lodge of Bushey Park (adjacent to Hampton Court) and the manor of Apscourt in Surrey. Flamsteed, who drew malicious pleasure from the news, valued the house and lands at £20,000, a small fortune with the additional £5000. Ambiguities cloud the picture, but it is impossible not to accept that Catherine Barton was Halifax's mistress and that Newton must have been aware of the relationship. In the nineteenth century this information was unacceptable and was therefore denied (but in any case it has no bearing on Newton's scientific achievement).

Newton's career at the mint brought him back to reality. It did not take him long to understand the true state of affairs. Historically the warden had been the institution's highest officer but in fact, after a reorganization in 1666, the master (that is, the master coiner) had assumed that role. He controlled the accounts—even the warden received his salary through the hands of the master. Not only was his salary higher than the warden's, but he received a set payment, or 'profit', from each pound of precious metal coined. When Newton arrived Thomas Neale was the master. He treated the position as a sinecure so that Newton, who had been offered a sinecure, ended up shouldering many of his duties. Not only did Neale receive a salary of £500 to Newton's £400, but his profits from the recoinage, to which he contributed almost nothing, mounted above £22,000. Almost at once after he assumed office in the mint, Newton, who knew that Neale was in poor health, began to familiarize himself in detail with all its operations. No warden had ever before become master but when Neale obliged by dying at the end of 1699 the appointment immediately went to Newton, and this despite the fact that his patron Montagu had fallen from power. He took office on 25 December, his birthday. The don from Cambridge had learned how to swim in the dangerous waters of London.

During more than twenty-seven years as master Newton's average income was about £1650, though it varied widely from year to year. In 1701, a year of heavy coinage, his income reached nearly £3500 and he decided to resign his fellowship and chair in Cambridge. The War of the Spanish Succession quickly brought coinage to a standstill; his total profit from coining in 1703 was £13. With peace, the pace picked up again. In 1715 his total income came to £4250, and in years of peace he averaged about £2250. Very few governmental positions carried a salary higher than that, and Newton died a wealthy man.

For all the trauma the recoinage imposed on English society, silver was allowed to remain undervalued so that it quickly disappeared to be sold abroad as bullion. The shortage of silver coins became a serious inconvenience, and over the years the Treasury called upon the master of the mint to offer advice. Papers with titles such as 'Observations upon the valuation of gold and silver in proportion to one another' urged that the value of the guinea, a gold coin, be set significantly lower than its current rate of 21*s*. 6*d*. Eventually a government which ignored his advice longer than it should have, and then acted only half way, lowered the value to 21*s*.; what proved to be the permanent value and later the definition of the guinea is a small fraction of Newton's legacy.

President of the Royal Society

The Royal Society hardly figured in Newton's activities during the first seven years of his residence in London. The society stood at a low ebb, with leadership lacking, interest declining, and finances in chaos. The death of Robert Hooke in March 1703 removed an obstacle to Newton's active participation, and at the annual meeting on 30 November of that same year the Royal Society looked to a real natural philosopher for leadership and elected Newton as president. He continued as president, dominating the society's affairs, until his death.

Master of the mint and president of the Royal Society: Newton had moved a long way from the reclusive don who had isolated himself in his chamber in Trinity College, and Queen Anne added a further plum to his pudding in 1705. Halifax, who needed a block of supporters in the House of Commons, persuaded Newton to stand once more for Cambridge in the election that year. This was the election that provoked the noisy demonstrations by students against nonconformity and determined Newton, who was not elected, never to run again. However, the election also had another effect. On the occasion of a visit by Queen Anne to the university Halifax organized what was really a campaign rally in which he received an honorary doctorate, and his brother was knighted along with Halifax's candidate for parliament. Thus Newton became Sir Isaac Newton, the first scientist in Europe so honoured (whatever the political calculations that lay behind it).

To the Royal Society, Newton brought the same organizational talent that he had brought to the mint, and the same inability to ignore an obligation. Whereas the previous two presidents, prominent political figures, had attended a total of three meetings during the space of eight years, Newton presided at more than three out of four, and during the twenty years that followed 1703, until age imposed constraints, failed to preside at only three sessions of the council. His first concern was the content of meetings. To the first one over which he presided he brought along the scientific instrument maker, Francis Hauksbee, to perform experiments with his air pump. Hauksbee became a fixture. His experiments, first with the air pump and then on electricity and capillary action, published as *Physico-Mechanical Experiments* in 1709, elevated him to the status of a well-known scientist in his own right, and some of his discoveries, especially those related to electricity, influenced Newton. When Hauksbee died in 1713, Newton found an equally capable replacement in J. T. Desaguliers, a Huguenot refugee.

Likewise the finances of the society revived under Newton's administration. A society on the verge of bankruptcy in 1703 was able to purchase a home of its own in 1710. Since its establishment, the society had met in Gresham College. In the summer of 1710 a house in Crane Court off Fleet Street, owned by the physician Edward Browne, son of Sir Thomas Browne, came up for sale. Newton moved quickly so that the society could buy it for £1450. There were additional expenses of £710 for repairs and renovation, but the society had only £550 in ready cash; gifts poured in, however, and in less than six years the society had retired all the debt and was able to invest an additional bequest in bank annuities.

The Royal Society provided a setting for Newton's second major publication, *Opticks*. On 16 February 1704, two and a half months after he assumed the chair, Newton presented the book to the society. The overwhelming majority of its content dated from the late 1660s and early 1670s: in the 1690s he had rewritten his optical lectures of 1669–72 into book one; with the exception of the section on coloured phenomena in thick transparent plates, work done during 1689, even the prose of book two repeated a paper written in 1672. What was new were the 'queries' that formed the bulk of book three. There were sixteen in the first edition, to which he added seven more (those numbered 25 to 31 in current editions of *Opticks*) in the Latin edition of 1706. In the twenty-three queries of 1704–6 Newton offered a general statement, couched in rhetorical questions that demanded affirmative answers, of his programme in science:

> Have not the small Particles of Bodies certain Powers, Virtues, or Forces, by which they act at a distance, not only upon the Rays of Light for reflecting, refracting, and inflecting them, but also upon one another for producing a great Part of the Phaemonema of Nature [the final

query began]? For it's well known, that Bodies act one upon another by the Attractions of Gravity, Magnetism, and Electricity; and these Instances shew the Tenor and Course of Nature, and make it not improbable but that there may be more attractive Powers than these. For Nature is very constant and conformable to her self.

When he wrote these words, and others to the same effect, the eight queries (numbered 17 to 24 in current editions) asserting the existence of a universal ether that offered quasi-mechanistic accounts of these forces did not exist. Newton added them to the second English edition as age brought on timidity. The *Opticks* of 1704–6, generalizing the message of the *Principia*, issued a manifesto of his new programme. The mechanical philosophy with its imagined mechanisms of imagined particles was out. Forces mathematically defined, explaining observed motions in a quantitatively precise manner, formed the centrepiece of Newtonian science. Modern science has shaped itself accordingly.

The same *Opticks* also contained at the end the first full papers in mathematics that Newton made public. He had been aware of Leibniz's publication of his differential calculus and of the renown it had brought him. Although word circulated about the Newtonian method of fluxions, the manuscripts of which only a small number studied, the only text expounding it in print was a brief précis of 'De quadratura' in John Wallis's *Opera.* 'Tractatus de quadratura curvarum' ('A treatise on the quadrature of curves') changed that situation, and 'Enumeratio linearum tertii ordinis' ('Enumeration of lines of the third order'), though not on fluxions, further displayed his mathematical prowess. They may be looked upon as a preliminary salvo in the battle that would be joined with Leibniz over priority in the discovery of the calculus.

A new confrontation with Flamsteed also arose from Newton's office of president. The failure with lunar theory still rankled. Newton was convinced that Flamsteed had a trove of observations that would enable him to achieve success. Therefore, only a few months after he assumed the society's chair, he went down to Greenwich, full of pretended benevolence, asking if he might recommend the publication of Flamsteed's observations to Prince George, the consort of Queen

Anne. There were few things that Flamsteed wanted more, and the prince's agreement to finance the publication was immediately gained. In the process Newton also succeeded in getting the project into his own hands. The group of referees in charge of publication, which he headed, paid no attention to Flamsteed's desires or advice and proceeded towards the publication of the data that Newton most wanted. It did not take long for the seemingly friendly relations at the beginning of the project to degenerate into outright hostility, and there they remained until Prince George died near the end of 1708, bringing publication to a temporary halt before it had progressed very far.

Newton was unwilling to let the matter stand. In 1710, as he embarked on a second edition of the *Principia* in which he would perforce deal again with lunar theory, he arranged for the queen to appoint the Royal Society as visitor, that is, supervisor, of the observatory, and from this position of power he again set out to publish Flamsteed's catalogue of the fixed stars and selected other observations, with expenses to be borne by the Treasury. This time there was no pretence of co-operation—tense antagonism reigned from the beginning. Newton had an imperfect copy of the catalogue from the earlier episode and, as visitor, forced Flamsteed to hand over observations from which Edmond Halley was able to fill in the lacunae. Thus, in 1712, *Historia coelestis*, with a shameful preface that denigrated Flamsteed, issued from the press. As he completed edition two of the *Principia*, Newton struck out fifteen references to Flamsteed, but his reliance on Flamsteed's observations of the great comet prevented him from reducing the astronomer royal to nonexistence.

This was not yet the end of the affair. In 1714 the death of Anne led to the fall of the tory government and the return of the whigs. When Halifax died the following year Newton's patron disappeared, whereas Flamsteed had one in the lord chamberlain, the duke of Bolton. Through Bolton, Flamsteed obtained an order that all unsold copies of the *Historia coelestis*, about 300, be delivered to him. Separating out some pages, he consigned the rest, including the catalogue of fixed stars, to the flames. Already he had begun, at his own expense, to print the catalogue and his observations in the form he had always wanted. The pages he saved from Newton's edition, his early observations before the construction of his great mural arc in 1689, became volume one in his edition. When he died in 1719, he had nearly completed volume two, the observations with the mural arc. His widow and two former assistants oversaw the printing of volume three, mostly the catalogue of the fixed stars. In 1725 *Historica coelestis Britannica* appeared, a landmark in the history of observational astronomy, and a monument to Newton's failure to reduce Flamsteed to subservience.

Revisions and disputes

6

Revising *Principia*

The climax of the struggle with John Flamsteed coincided with a number of other crises in Newton's life. On 5 May 1711 an angry confrontation with Craven Peyton, warden of the mint, signalled the deterioration of their relationship. Tension continued at a high level until the death of Anne in 1714 and the fall of the tory government, with which Peyton was allied, led to the appointment of a new warden. During the same period, beginning in this case in 1709, Newton engaged himself in a second edition of the *Principia*, which taxed his time and energy until its completion in 1713. And with the Royal Society's receipt, on 22 March 1711, of a letter from Leibniz, the long smouldering dispute between him and Newton over priority in the invention of the calculus finally burst into open flame. The seven years from 1709 to 1716, when Leibniz died, were among the most stressful in Newton's life.

Newton had been thinking about a second edition of his masterpiece almost from the day it first appeared, but the specific timing of the edition owed less to Newton than to Richard Bentley, now master of Trinity, who was determined to woo Newton's support for his efforts to reform and rejuvenate the college. Bentley arranged the edition and functioned as its publisher. He appointed one of his young protégés, Roger Cotes, only twenty-seven years old but already Plumian professor of astronomy at Cambridge, as editor, and in 1709 Newton and Cotes got down to work. Bentley's choice could hardly have been more fortunate. Despite Newton's early, repeated suggestions that he not give himself too much trouble, Cotes insisted on going over the copy with a fine-tooth comb, and in the end succeeded in luring Newton into a serious discussion of his work. As a result the second edition of the *Principia* contained a number of major emendations that greatly enhanced its value. Today only historians of science refer to the first edition;

the *Principia* that has shaped Western scientific tradition is substantially the second edition, very modestly modified in a third edition shortly before Newton's death.

Initially the two men found little more than details to discuss, and by the middle of 1710 the edition had moved swiftly through book one and well into book two, not far short of two-thirds of the entire work, before they struck the issue of a fluid's resistance to motion through it. This was an important topic, the heart of Newton's argument against Cartesian philosophy. In the first edition it had rested on a dubious experimental basis, and difficulties continued to appear as Newton performed new experiments in London and fundamentally revised the theory. At the time when the confrontation with Flamsteed was boiling over, and when Newton's relationship with Peyton at the mint was in a state of acute tension while heavy coinage imposed additional strain, the press simply came to a halt until the problem was resolved. For all the effort, the section on the resistance of fluids remained the *Principia's* weakest part. Meanwhile, printing hardly advanced through the rest of 1710 and through 1711. By the end of 1711 the dispute with

Leibniz was beginning to expand to include their philosophical differences, and the new edition, important as it was in its own right, acquired additional urgency. Devoting himself to it seriously once more Newton, ably assisted by Cotes, drove it to completion without another major interruption.

At the beginning of book 3, a new 'rule of reasoning in philosophy' ('rule 3') flung down the experimental gauntlet in challenge to Leibniz's more speculative philosophy. Other changes late in book two and in book three emphasized a concomitant advantage of Newtonian philosophy: its ability to calculate natural phenomena, such as the speed of light and the rate of precession of the equinoxes, with quantitative precision. The two major additions to the work, a long preface by Cotes and a concluding 'General scholium' by Newton, both stressed the differences that separated Newtonian philosophy from its continental rival. The 'General scholium', one of Newton's last testimonies about his vision of the scientific enterprise, began with a repudiation of mechanical philosophies in general: 'The hypothesis of vortices is pressed with many difficulties', he asserted, difficulties that he proceeded to detail. Moreover, the order of the cosmos seemed incompatible with mechanical necessity and demanded not merely a creator at the beginning but his abiding presence. In one of his most quoted passages, which argued that mechanical causes could not explain gravity, Newton proclaimed anew the difference that separated his philosophy from speculative, mechanical ones:

> But hitherto I have not been able to discover the cause of those properties of gravity from phenomena, and I feign no hypotheses... And to us it is enough that gravity does really exist, and act according to the laws which we have explained, and abundantly serves to account for all the motions of the celestial bodies, and of our sea. (*Mathematical Principles*, 547)

As the edition neared completion there arose a final crisis. Nikolaus Bernoulli, the nephew of Johann Bernoulli, visited London, where he told Newton's friend Abraham De Moivre that his uncle had found an error in proposition 10 of book three. The error seemed to indicate that Newton did not understand second differentials, and Newton had no intention of allowing any such thing to appear in print, especially with the priority dispute in full blaze. In a *tour de force* for a man of nearly seventy, he located the source of the error and corrected it. Proposition 10 had long since been printed. A whole sheet had to be set anew, together with the last two pages on the previous sheet, which were glued onto the stub of the original, mute testimony to the late correction. Johann Bernoulli, Leibniz's supporter, who had intended to use the mistake as evidence that Newton's fluxional method was inferior to the differential calculus, was furious to find that Newton had not attributed the correction to him.

Battles with Leibniz

At the end of June 1713 the long labour reached its fulfilment and the second edition of *Principia* emerged from the press. Unfortunately the controversy with Leibniz had no similar termination. The conflict had long been simmering, since 1684 to be exact, when Leibniz first began to publish his calculus without mentioning the correspondence of 1676 by which he knew that Newton had a similar method. Newton grumbled about it but took no action. In 1699 he did co-operate in the publication, in volume three of Wallis's *Opera*, of his two letters to Leibniz in 1676 plus other letters that testified to his progress by the year 1673. That same year, 1699, in a mathematical tract, Fatio de Duillier, who had seen Newton's papers and heard his complaints, bluntly asserted the Englishman's priority and only a little less bluntly accused Leibniz of plagiarism. For his part, Leibniz made similar charges against Newton in reviews that he published anonymously. To one of those charges a young Oxford don, John Keill, a protégé of James Gregory, replied directly in a paper published in the *Philosophical Transactions* in 1708, again asserting Newton's priority and accusing Leibniz of plagiarism. When he eventually saw the paragraph in question, Leibniz wrote to the Royal Society, of which he was a member. His letter, received on 22 March 1711, demanded that Keill publicly recant.

Newton's papers contain an enormous record of his personal response to Leibniz's letter, as he went back over his own papers and reconstructed the order of events. At meetings of the society during April he recounted the history of his method. Newton had trained himself in empirical historical research in his theological investigations, and that capacity, with its command of factual details, he brought to the priority dispute. He appears to have participated in composing Keill's reply to Leibniz, and he certainly composed the covering letter from Hans Sloane, secretary of the Royal Society, that accompanied it.

Leibniz, in no way satisfied to see his demanded apology refused, replied in a letter received early in 1712. Distinguishing between Keill and Newton, expressing his regard for the latter (in contradiction to his anonymous reviews) but asserting his equal right as an independent discoverer, he threw himself on the society's sense of justice. Newton made an unexpected riposte by having the society appoint a committee to sit in judgement. The society, that is, Newton, liked to refer to it as a committee made up of gentlemen of several nations; in fact it was composed of Newton's followers plus Frederick Bonet, the minister in London of the king of Prussia, who allowed himself to be co-opted. Newton himself wrote the committee's report, drawing upon what was by then a year's research in his records. In a ringing condemnation of Leibniz and a full endorsement of Newton, the report insisted that there were not two independent methods but a single one, and cited the correspondence of 1676 together with Leibniz's access to Newton's papers on his visit to London in that year to convict him of plagiarism. The society went on to publish the report, together with the factual basis in the early correspondence that gave it weight, in a volume, *Commercium epistolicum D. Johannis Collins, et aliorum de analysi promota* ('The correspondence of the learned John Collins and others relating to the progress of analysis'), copies of which it sent about through the learned society of Europe. Newton put the volume together, of course, and he later reviewed it at great length, anonymously, in the *Philosophical Transactions*.

It was out of the question that so blatantly partial a publication as the *Commercium epistolicum* should settle the issue. Leibniz replied in his own way, avoiding the ground of historical details that Newton commanded and emphasizing instead the differences of the two methods and the superiority of his own. Proposition 10 of book two saw heavy duty. Like Newton, Leibniz tried to conceal his own role in the dispute, in his case by relying on anonymous publications and the testimony of others. The dispute refused to die or to confine itself to the principal figures. Nor was it confined to mathematics, as differences in natural philosophy and eventually theology crept in. In a letter to the princess of Wales, who had been his protector in Germany, Leibniz criticized Newton's published understanding of the relation of God to the creation. Out of the letter came a correspondence on related issues between Leibniz and Samuel Clarke, the leading theologian in England at the time and a dedicated Newtonian, who defended Newton's position and possibly had Newton's assistance in composing his part of the correspondence. The exchange went through five rounds, ten letters in all, each longer than its predecessor, and would no doubt have continued indefinitely, like the priority dispute, had Leibniz not died on 4 November 1716.

If the Clarke–Leibniz correspondence perforce came to an end, the priority dispute had generated too much heat and involved too many additional men, especially Johann Bernoulli on the one hand and the mathematicians John Keill and Brook Taylor on the other, to extinguish itself. For another six years letters passed back and forth and literary journals published attacks and replies. Eventually these too ran down, and at last silence, if not peace, reigned.

Return to theology

During the final decade of his life Newton spent some of his time attending to his scientific legacy. A second English edition of the *Opticks* in 1717 scarcely touched the text but added eight new queries, which partially compromised with orthodox mechanical philosophy by reasserting the existence of a cosmic ether, a concept he had consciously excluded from his philosophy since about 1680. The compromise was more apparent than real. If the new ether explained gravity and optical phenomena in a seemingly familiar manner, its structure of particles that repelled each other at a distance incorporated the very notion that mechanical philosophers found inadmissible. In 1721 a third English edition, which scarcely differed from the second, appeared. Meanwhile, a second Latin edition (1719), in the language of international scholarship, was more important. Together with two French editions (1720 and 1722) it carried Newton's conclusions in optics, which heretofore had not spread much beyond Britain, to the continent. Newton also participated in a 1726 third edition of the *Principia*, which did not begin to rival the second in the importance of its emendations.

 Newton also functioned as a scientific expert for the government. In 1714, after hearings in which Newton testified along with others, parliament offered a large monetary prize for a method of determining longitude at sea. A board of longitude, on which, inevitably, Newton sat, was established to judge submitted proposals. Until death delivered him, one of his duties was to read these proposals, most of them verging on the absurd.

It was theology rather than science, however, that dominated the consciousness of the ageing man. Some time between 1705 and 1710 he returned to the subject he had largely ignored for two decades, and theology formed the principal staple of Newton's intellectual life from that time until his death. Like his late efforts in science, those in theology confined themselves mostly to reshuffling earlier ideas. This he did at enormous length, so that an immense volume of manuscripts testifies to his activity. One theme unites much of it. Newton had become a prominent man of the world who did not intend to compromise his position by publicly espousing opinions that had passionately stirred an isolated young don in Cambridge. Much of his effort was devoted to laundering those opinions to obscure their radical thrust.

The most radical of Newton's theological endeavours had been his 'Theologiae gentilis origines philosophicae', which he transformed into the *Chronology of Ancient Kingdoms Amended*, as the manuscript published soon after his death was entitled. In 1716 an Italian visitor to England, Abbé Antonio Schinella Conti, both Newton's confidant and a familiar figure in the court, mentioned Newton's new principles of chronology to the princess of Wales, who immediately wanted to see the treatise. Newton had no intention of surrendering a manuscript he considered potentially damaging. Because he could not refuse a royal command, he hastily composed an 'Abstract', later called the 'Short chronology', which put the work in a shape, little more than a list of dates, which Newton deemed suitable for the princess's eyes.

The 'Abstract' had its own history. When Conti left England he carried with him a copy, which he showed about Paris. French scholars in chronology rejected its apparently arbitrary truncation of ancient history, and one of them arranged to publish a translation of it together with a refutation.

 Thus Newton was drawn into another controversy, which was not as bitter and not as prolonged as the priority dispute but was sufficiently sharp nevertheless. When he died his heirs found the completed manuscript of the *Chronology*, which they immediately sold to a publisher for £350.

Newton also returned to his study of the prophecies. Here too he obscured the radical, Arian thrust of his early interpretation, partly by placing Daniel rather than Revelation at the centre so that the rise of trinitarianism, the object of his odium, was less prominent, and in general by converting the work into a set of rambling commentaries that had no obvious point. At his death he left a completed treatise and a newer, still incomplete one. Advisers to his heirs melded the two into one and added three additional chapters that Newton had not considered part of either; eventually *Observations upon the Prophecies of Daniel, and the Apocalypse of St. John* (1733) appeared.

Other theological manuscripts, which did not find publication in the eighteenth century and remained largely unknown until the mid-twentieth, were more reminiscent of the earlier stance, though here too Newton couched them in a milder idiom. One of these, which survives in multiple versions, bore the name 'Irenicum'. On the surface it presented a programme of common beliefs on which all Christians could agree, but this programme dispensed with all the distinctive doctrines of Christianity and reduced religion to two principles, love of God and love of neighbour. 'This was the religion of the sons of Noah established by Moses & Christ still in force' (King's Cam., Keynes MS 3, 5–7). 'Thus you see there is but one law for all nations', he stated in another paper, 'the law of righteousness & charity dictated to the Christians by Christ to the Jews by Moses & to all mankind by the light of reason & by this law all men are to be judged at the last day' (Keynes MS 7, 2–3). Only within a very confined circle did Newton discuss such opinions. Publicly he performed as a trustee of the chapel on Golden Square, and as a member of the commission established by parliament to oversee the construction of fifty new churches in the expanding suburbs of London, and of the commission to supervise the completion of St Paul's Cathedral.

'Let Mortals rejoice'

In June 1717 a young man who figured prominently in the rest of Newton's life read a paper to the Royal Society on the site of the Roman city Carteia, in Spain, and two months later he gave Catherine Barton married respectability. Not only was John Conduitt the husband of Newton's favourite niece, he was also an unabashed hero worshipper of Newton himself. To his determination to assemble a record of the great man is owed many of the details known about Newton, and to him also is owed the preservation of Newton's papers. Through negotiation with the other heirs the Conduitts obtained possession of the papers, and when their only child, Catherine, married John Wallop, Viscount Lymington, the papers passed to the Portsmouth family, which later donated the bulk of them to Cambridge University Library. Conduitt himself, who acted unofficially as Newton's deputy at the mint during the final two years, succeeded him as master.

Conduitt and William Stukeley set down much of the accepted picture of Newton the man. Conduitt reported that he was of medium stature, though Thomas Hearne called him short. Both agreed that he was plump, but this must have come with age for his early portraits show no such thing. Count Luigi Ferdinando Marsigli, who met him in 1723, was struck by his small wizened figure, and reported that, like many English scholars, he was unable to converse fluently in Latin. In general he was silent, a man who spoke little in company. Humphrey Newton (no relation), his amanuensis for five years in Cambridge, saw him laugh only once; Stukeley, trying unsuccessfully to mitigate the image of constant gravity, said that he could easily be brought to smile if not to laugh.

The first portrait of Newton (1689), by Godfrey Kneller, shows him at the height of his powers, immediately after the *Principia*. Kneller painted at least three more portraits, the last when Newton was about eighty. During his London years portraiture became something of an obsession with Newton. Beyond those by Kneller, there were two by James Thornhill, three by John Vanderbank, and at least seven others, one or two of which may be copies, plus a sketch by Stukeley, a medal by the engraver at the mint, and two ivory busts (nearly identical with each other) and three ivory plaques by David Le Marchand.

During his last five years Newton's health steadily deteriorated. There was a serious illness during the spring of 1722, and a more serious one in January of 1725. On 2 March 1727 he presided over a meeting of the Royal Society, was exhilarated by the experience, but overtaxed himself and collapsed the following day. He never recovered. In his most significant act as he lay dying, he refused to take the sacrament of the Church of England. On 15 March he seemed somewhat better but immediately declined again and died on 20 March 1727.

Newton's death did not go unnoticed. He lay in state in the Jerusalem Chamber in Westminster Abbey, and, with the pall borne by the lord chancellor and five high-ranking aristocrats, was interred on 28 March in a prominent position in the nave, where in 1731 his heirs erected an extravagant monument. Few, however, have found the final lines of the inscription extravagant: 'Let Mortals rejoice That there has existed such and so great an Ornament to the Human Race'. There has never since been a time when Newton was not considered either the greatest scientist who ever lived or one of a tiny handful of the greatest. His *Principia* marked the culmination of the scientific revolution, which ushered in modern science, and through its legacy the work may have done more to shape the modern world than any other ever published.

Sources

The correspondence of Isaac Newton, ed. H. W. Turnbull and others, 7 vols. (1959–77) · *The mathematical papers of Isaac Newton*, ed. D. T. Whiteside, 8 vols. (1967–80) · *The optical papers of Isaac Newton*, ed. A. E. Shapiro (1984–) · *Sir Isaac Newton's Mathematical principles of natural philosophy and his system of the world*, trans. A. Motte, rev. F. Cajori, 2 vols. (1934) · *Isaac Newton's Philosophiae naturalis principia mathematica*, ed. A. Koyré, I. B. Cohen, and A. Whitman, 2 vols. (1972) · I. Newton, *Opticks*, 4th edn (1952) · *Unpublished scientific papers of Isaac Newton*, ed. A. R. Hall and M. B. Hall (1962) · *Isaac Newton's papers and letters on natural philosophy and related documents*, ed. I. B. Cohen (1958) · The National Archive, Mint MSS, 19.1–5 · King's Cam., Keynes MSS · Jewish National and University Library, Jerusalem, Yahuda MSS · CUL, Portsmouth MSS, Add. MSS 3958–4006 · MIT Cam., Dibner Institute, Babson MSS · Smithsonian Institution, Washington, DC, Burndy MSS, Dibner collection · Bodleian Library, Oxford, MSS New College, 361.1–4 · I. B. Cohen, *The Newtonian revolution* (1980) · I. B. Cohen, *Introduction to Newton's 'Principia'* (1971) · A. Koyré, *Newtonian studies* (1965) · J. W. Herivel, *The background to Newton's 'Principia'* (1965) · B. J. T. Dobbs, *The Janus faces of genius: the role of alchemy in Newton's thought* (1991) · A. E. Shapiro, *Fits, passions, paroxysms: physics, method and chemistry and Newton's theories of colored bodies and fits of easy reflection* (1993) · F. E. Manuel, *A portrait of Isaac Newton* (1968) · R. S. Westfall, *Never at rest: a biography of Isaac*

Newton (1980) · K. A. Baird, 'Some influences upon the young Isaac Newton', *Notes and Records of the Royal Society*, 41 (1986–7), 169–79 · D. B. Meli, *Equivalence and priority: Newton versus Leibniz* (1993) · J. B. Brackenridge, *The key to Newton's dynamics* (1995) · S. Chandrasekhar, *Newton's 'Principia' for the common reader* (1995) · F. De Gandt, *Force and geometry in Newton's 'Principia'*, trans. C. Wilson (1995) · B. J. T. Dobbs, *The foundations of Newton's alchemy, or, The hunting of the greene lyon* (1975) · K. Figala, *Newton as alchemist* (1979) · C. W. Foster, 'Sir Isaac Newton's family', *Reports and Papers of the Architectural Societies of the County of Lincoln, County of York, Archdeaconries of Northampton and Oakham, and County of Leicester*, 39, pt 1 (1928), 1–62 · A. R. Hall, *Philosophers at war: the quarrel between Newton and Leibniz* (1980) · W. G. Hiscock, *David Gregory, Isaac Newton and their circle* (1937) · A. Koyré, 'A documentary history of the problem of fall from Kepler to Newton', *Transactions of the American Philosophical Society*, new ser., 45 (1955), 329–95 · D. McKie and G. R. De Beer, 'Newton's apple', *Notes and Records of the Royal Society*, 9 (1951–2), 46–54, 333–5

Index

T

W

The finest scholarship on the greatest people...

Many leading biographers and scholars have contributed articles on the most influential figures in British history: for example, Paul Addison on Winston Churchill, Patrick Collinson on Elizabeth I, Lyndall Gordon on Virginia Woolf, Christopher Ricks on Alfred Tennyson, Frank Barlow on Thomas Becket, Fiona MacCarthy on William Morris, Roy Jenkins on Harold Wilson.

'Paul Addison's Churchill *... is a miniature masterpiece.'*

Piers Brendon, *The Independent*

Every story fascinates...

The *Oxford DNB* contains stories of courage, malice, romance, dedication, ambition, and comedy, capturing the diversity and delights of human conduct. Discover the Irish bishop who was also an accomplished boomerang thrower, the historian who insisted on having 'Not Yours' inscribed on the inside of his hats, and the story of the philanthropist and friend of Dickens Angela Burdett-Coutts, who defied convention by proposing to the Duke of Wellington when he was seventy-seven and she was just thirty. He turned her down.

'Every story fascinates. The new ODNB will enrich your life, and the national life.'

Matthew Parris, *The Spectator*

At 60,000 pages in 60 volumes, the *Oxford Dictionary of National Biography* is one of the largest single works ever printed in English.

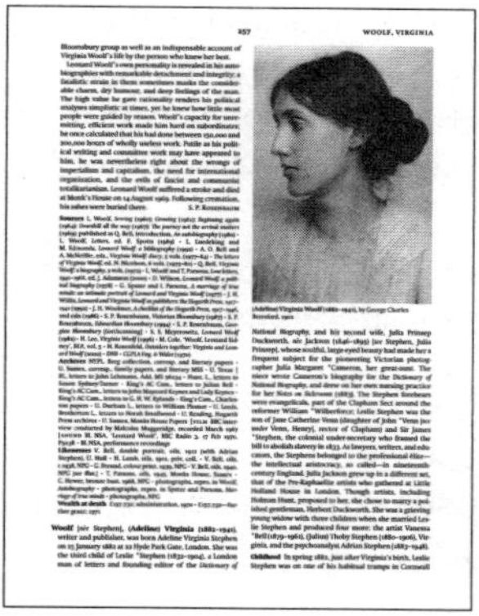

The award-winning online edition of the *Oxford DNB* makes it easy to explore the dictionary with great speed and ease. It also provides regular updates of new lives and topical features.

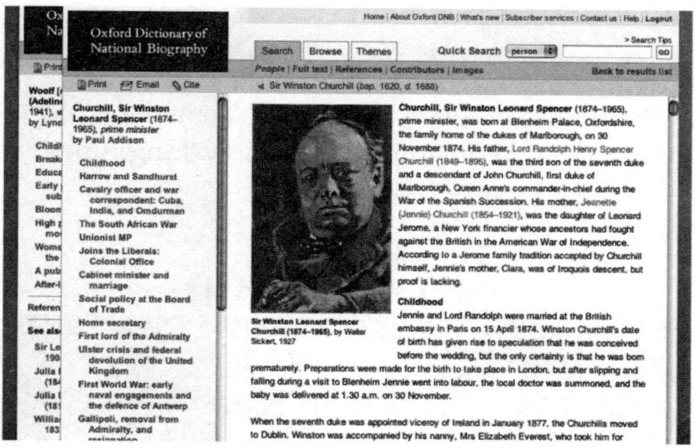

www.oxforddnb.com